Meet LEFTY ST. JAMES

Written by
Susan Sangiamo

Illustrated by
Adam McHeffey

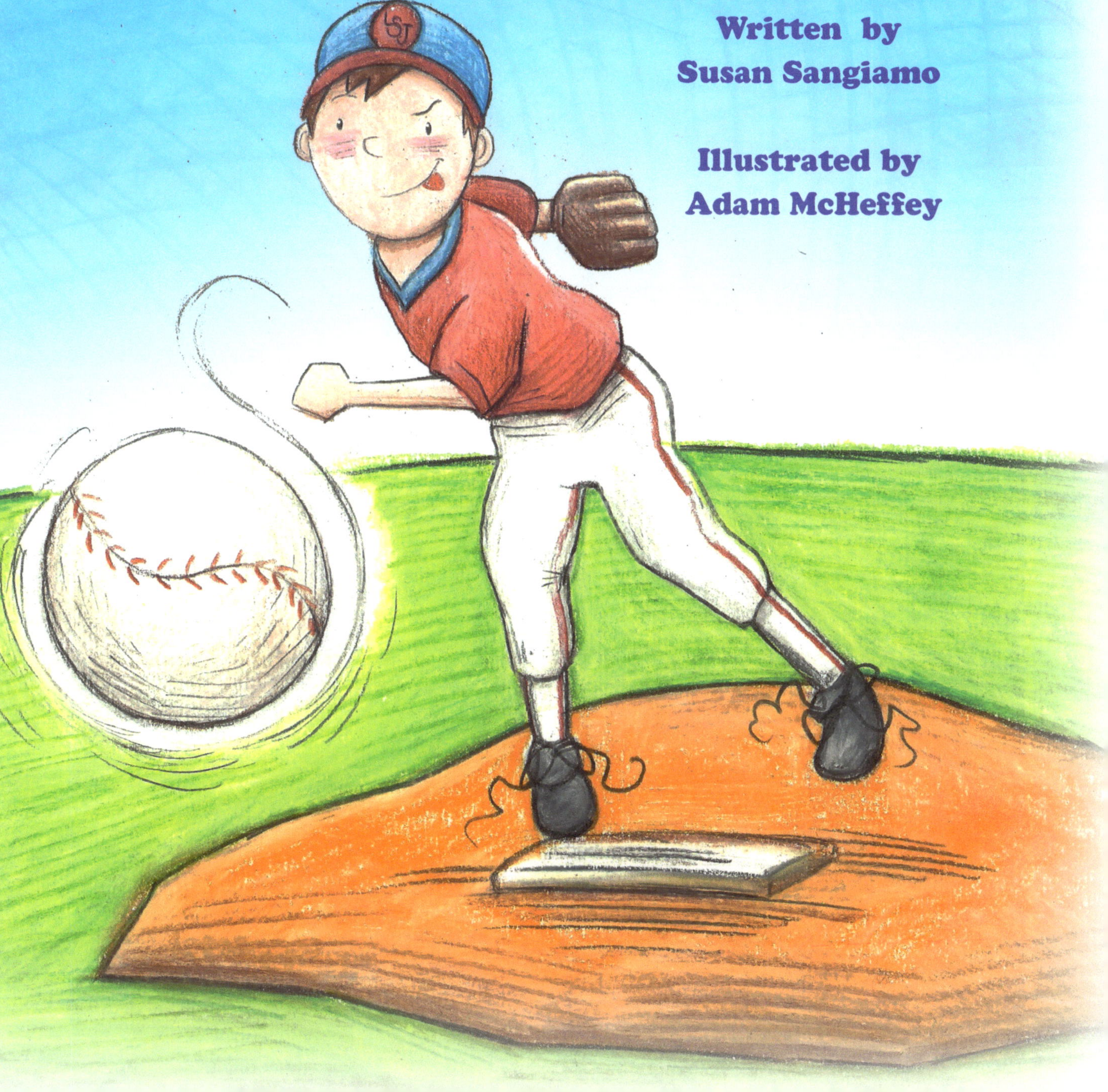

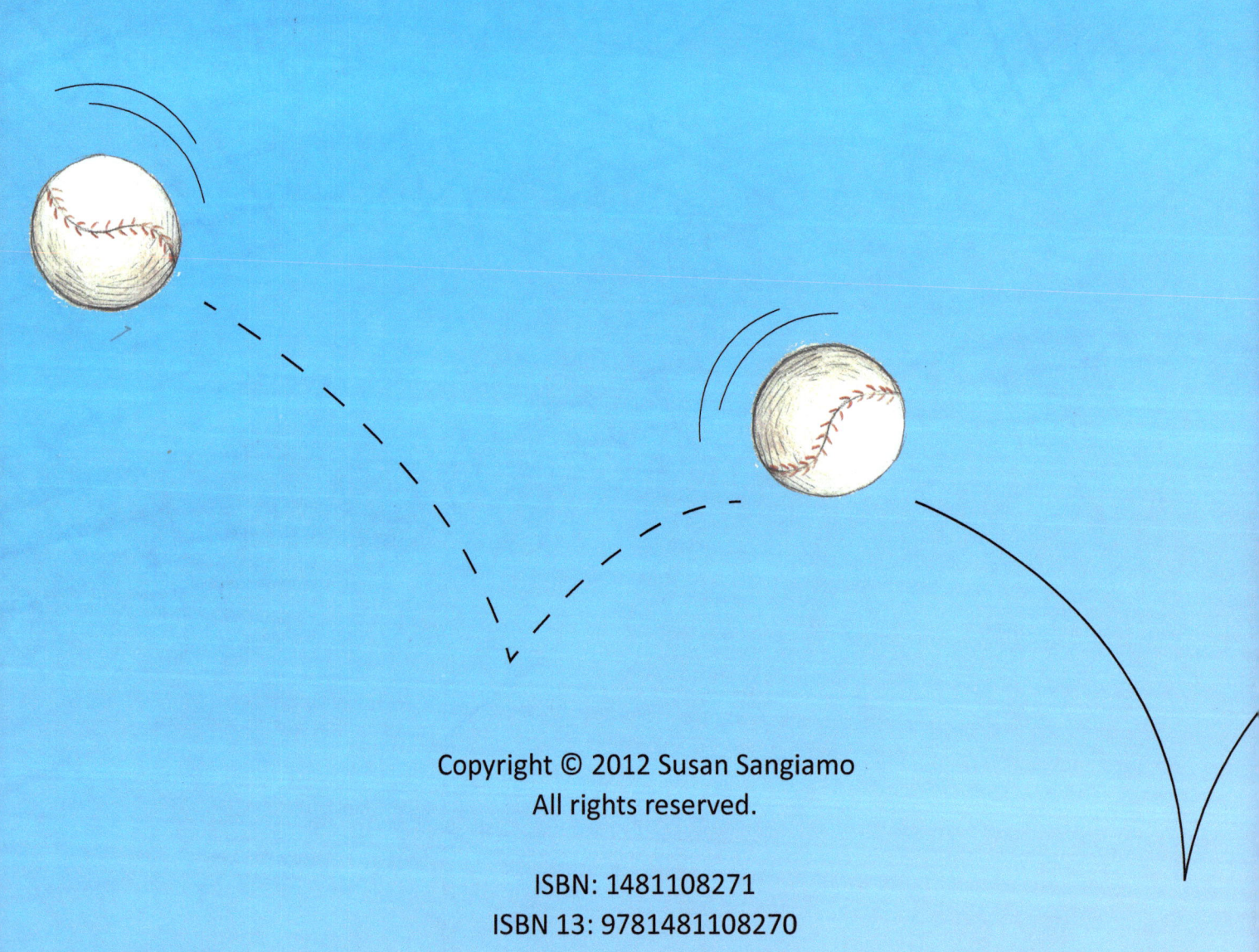

ISBN: 1481108271
ISBN 13: 9781481108270

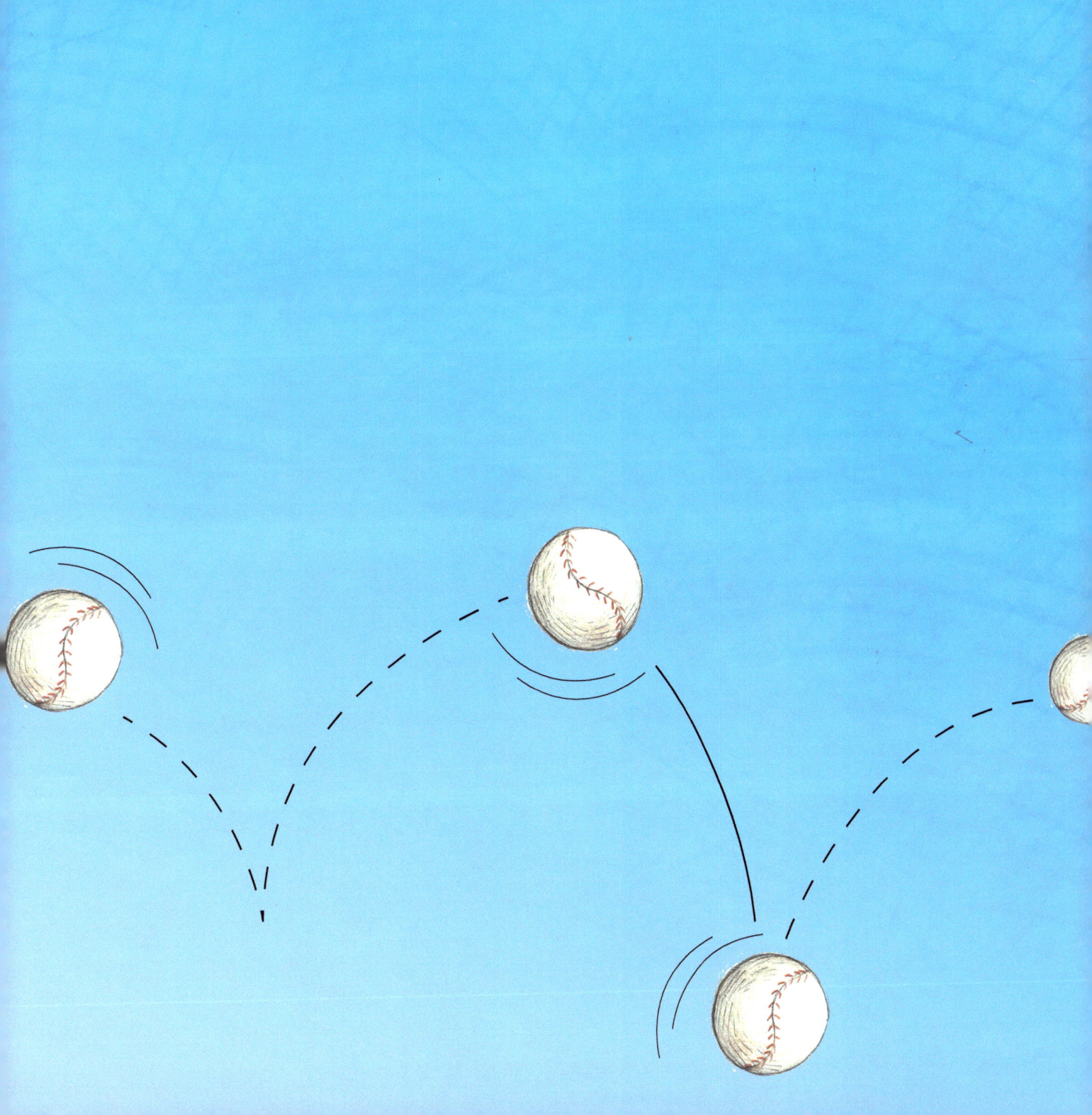

Dedication

To my son Michael, for his love of
baseball, his hard work, and for
inspiring me to fulfill my dream.
Love you forever.

For all students who will have as much fun
reading this book, as I had while writing it.

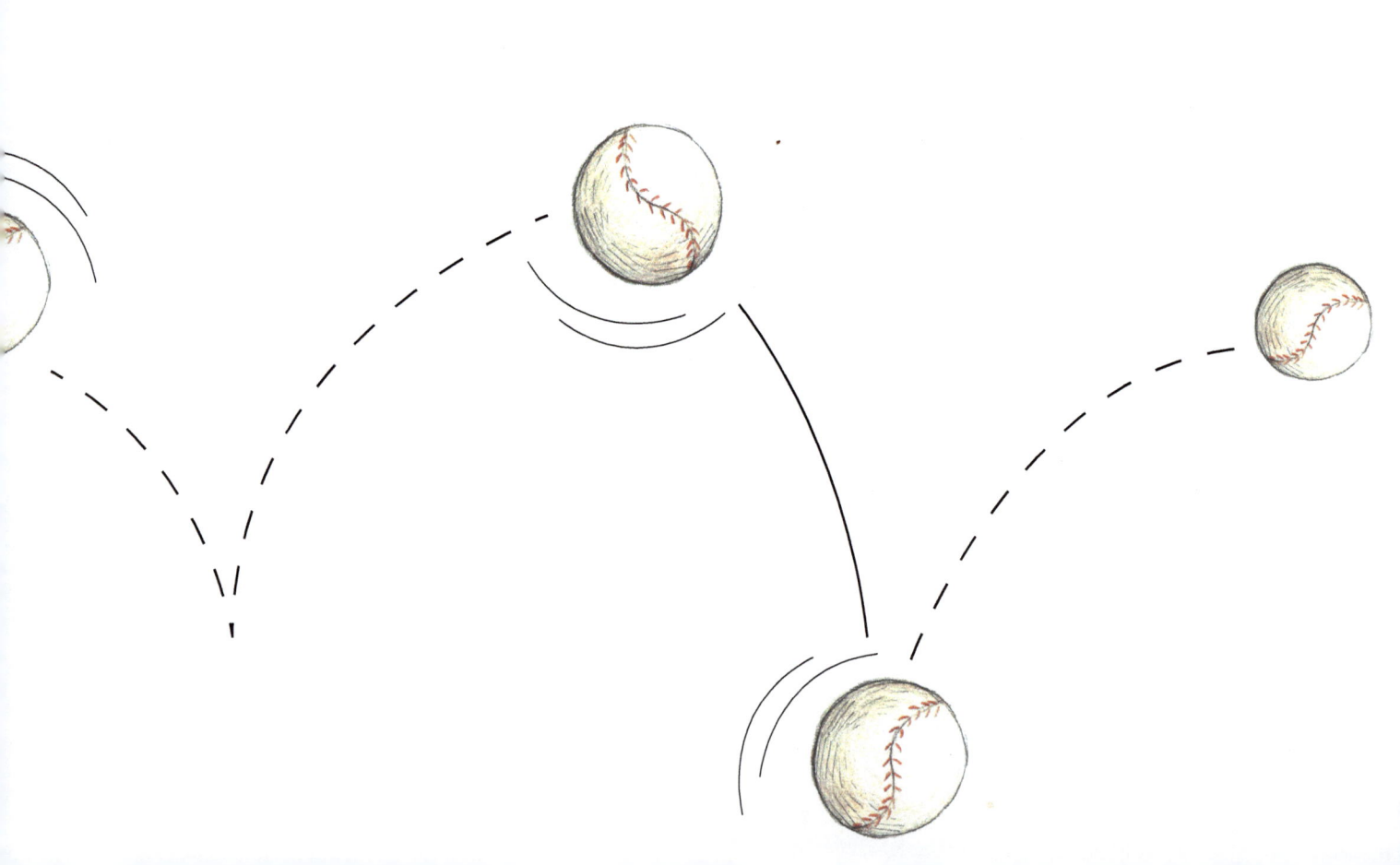

He'd practice each day to perfect his style...
He got really good after a while.

2

He threw really hard from the mound
with such ease...

That sometimes the catcher would fall to
his knees.

Being tall for his age, made lots of things easy...
But when Lefty did **math**, he got a bit queasy!

Lefty's mom tried to tell him how math can relate...
Just imagine you're standing up at home plate!

**Whether a base hit, a triple or even a homerun...
When you estimate the distance, it can be really fun!**

220'

165'

After throwing the ball, Lefty's pitch was complete...
From the mound to home plate, was how many feet?

When Lefty began to think of fractions
and rounding...
He thought, "Wow, the connection to
baseball is really astounding!"

26.5 SEC. ≈ 27 SEC.

So when it came to batting
averages, Lefty thought about...
How division was needed to figure them out.

When the coach had the team running
laps round and round...
The circumference was calculated and
easily found.

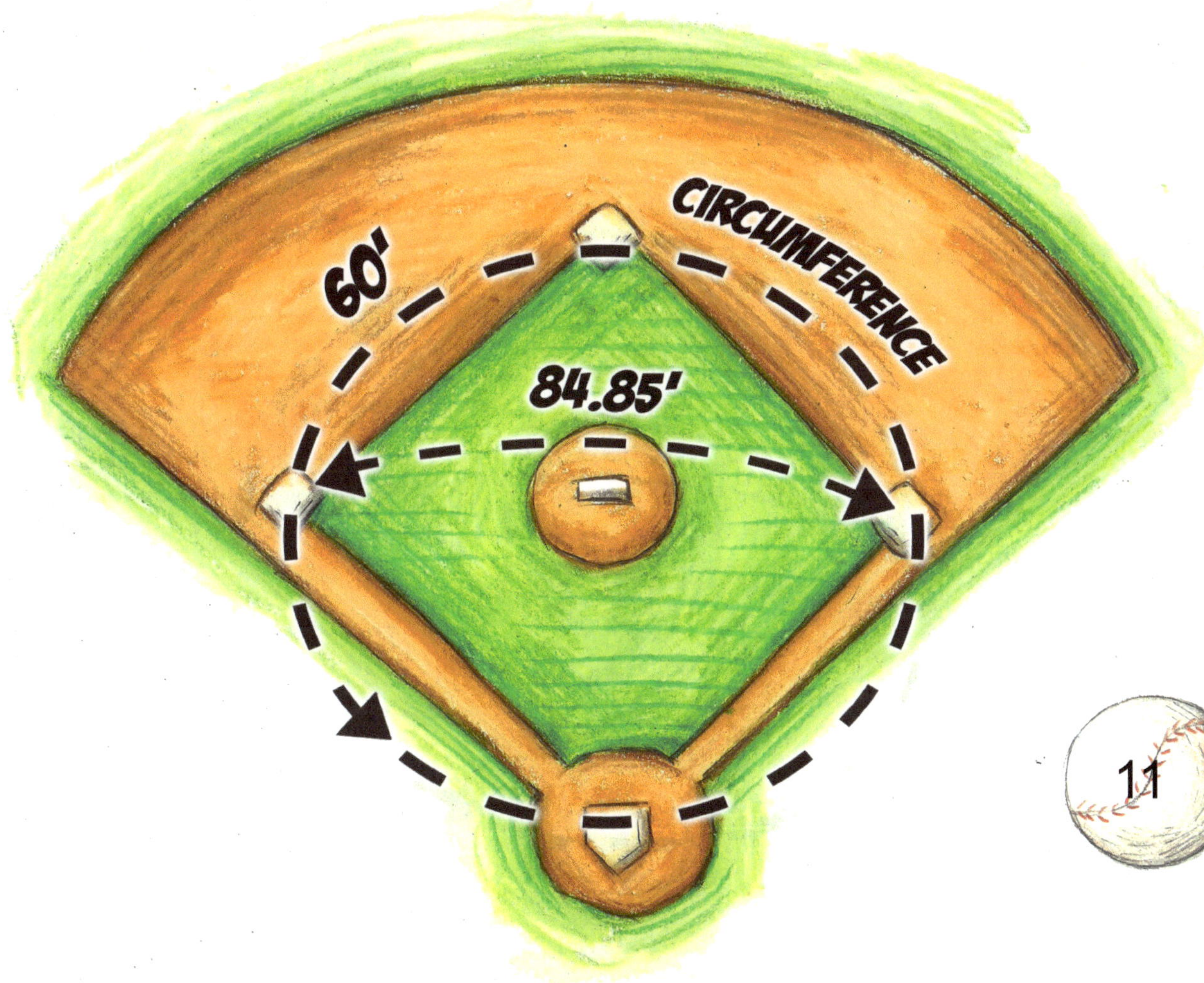

11

Math made better sense with this valuable lesson...
Remembering math concepts would keep Lefty from guessin'.

$$C = 2\tfrac{7}{8}'' \times$$

$$C = 9''$$

In his field of dreams,
where Lefty stood tall...
This is where he imagined playing pro ball.

13

That one day when he was big and was all done with school...
He remembered how baseball and math were both cool.

14

So, if you have a dream like Lefty one day...
Never let math ever get in your way.

15

Always strive for your best and hold
your head high...
Try making connections,
just give it a try!

16

It worked for Lefty!

17